Grade 3 · Unit 2

Inspire Science

Life Cycles and Traits

FRONT COVER: (t)Pavliha/iStock/Getty Images, (b)Olga P/Shutterstock;
BACK COVER: Pavliha/iStock/Getty Images; **SPINE** Pavliha/iStock/Getty Images

Mheducation.com/prek-12

STEM McGraw-Hill is committed to providing instructional materials in Science, Technology, Engineering, and Mathematics (STEM) that give all students a solid foundation, one that prepares them for college and careers in the 21st century.

Send all inquiries to:
McGraw-Hill Education
8787 Orion Place
Columbus, OH 43240

ISBN: 978-0-07-699627-8
MHID: 0-07-699627-1

Printed in the United States of America.

6 7 8 9 10 11 LWI 26 25 24 23 22 21

Table of Contents
Unit 2: Life Cycles and Traits

Plants

Animals

Plants

ENCOUNTER THE PHENOMENON

How do seeds grow?

Helicopter

GO ONLINE

Check out *Helicopter* to see the phenomenon in action.

Talk About It

Look at the photo and watch the video of the helicopter seed. What questions do you have about the phenomenon? Talk about your observations with a partner.

Did You Know?

About 2,000 different types of plants are used by humans as food.

STEM Module Project Launch

Science Challenge

Growing Plants

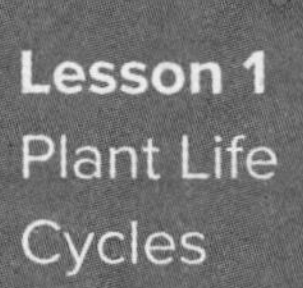

Lesson 1
Plant Life Cycles

You have been hired as a botanist to design a school garden. The garden should include three types of plants that are easy to grow and can be harvested in less than 40 days. You will need to provide information on the life cycle, traits, growth requirement, and care instruction for each plant.

What do you think you need to know to design a garden?

Botanists apply their knowledge of plants and their needs to keep plants healthy and growing.

POPPY
Park Ranger

Lesson 2
Plant Traits

STEM Module Project

Plan and Complete the Science Challenge Use what you learn throughout the module to complete the design of your garden.

PAGE KEELEY SCIENCE PROBES

LESSON 1 LAUNCH

Plant Life Cycles

Group A	Group B	Group C	Group D
cherry tree	pine tree	peas	potato
apple tree	spruce tree	corn	onion
strawberry plant	western larch	nuts	fern

In the table above, the plants are in groups.

How many of the groups contain examples of plants that have a life cycle? Circle the answer that best matches your thinking.

A. *None of the groups*

B. *1 group*

C. *2 groups*

D. *3 groups*

E. *All of the groups*

Explain your thinking. How did you decide if a plant had a life cycle?

You will revisit the Page Keeley Science Probe later in the lesson.

LESSON 1

Plant Life Cycles

ENCOUNTER
THE PHENOMENON

How is the dandelion changing?

GO ONLINE

Check out *Growing Seed* to see the phenomenon in action.

Talk About It

Look at the photo and watch the video of the growing seed. What questions do you have about the phenomenon? Talk about them with a partner. Record your thoughts below.

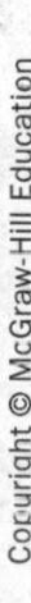

Did You Know?

The word *dandelion* comes from the French *dent de lion,* which means "lion's tooth."

INQUIRY ACTIVITY

Hands On

Seed Growth

You watched a seed grow in the video. Have you ever grown a plant from a seed?

Make a Prediction How will a lima bean seed and radish seed grow differently?

____

____

____

____

Materials

- 2 resealable plastic bags
- 2 paper towels
- lima bean seeds
- radish seeds
- water

Carry Out an Investigation

1. Label one bag "lima bean seeds." Fold a sheet of paper towel so that it will fit inside the bag.
2. Wet the paper towel with water. Place the wet paper towel in a bag, and lay the bag flat.
3. Place some lima bean seeds on the paper towel.
4. Repeat steps 1–3 with radish seeds.
5. Seal each of the bags, and place them next to each other in a safe, warm place.
6. **Record Data** Without opening the bag, observe the seeds daily. Make sure the paper towels stay moist. Record your observations about how the seeds change.

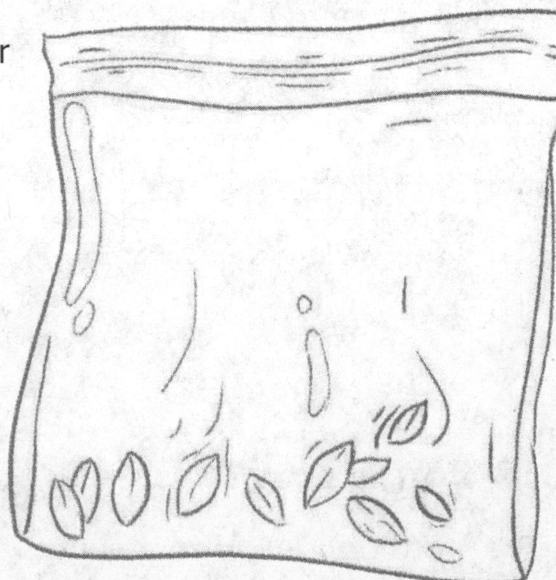

Seed Observations		
	Lima Bean Seeds	Radish Seeds
Day 1		
Day 2		
Day 3		
Day 4		
Day 5		
Day 6		
Day 7		
Day 8		
Day 9		
Day 10		

Communicate Information

7. What are some similarities and differences between how these two seeds sprout?

Talk About It

Did your findings support your prediction? Explain.

VOCABULARY

Look for these words as you read:

germinate

life cycle

pollination

reproduce

From Seed to Plant

Scientists estimate that there are about 400,000 different species of plants on Earth. Plants come in many shapes and sizes. Although plants can look very different, they share many of the same characteristics and basic structures, or parts. Most have roots, stems, and leaves. Some plants have flowers, fruits, and seeds. Other plants, such as pine trees, have cones that make seeds. Flowers, seeds, and cones help plants reproduce.

When conditions are right, a seed will **germinate**. To germinate means to grow from a seed to a young plant. A seed will germinate when the temperature is right and the soil is the right condition.

1. Based on what you observed in the *Seed Growth* activity, what are some common patterns in germination?

____

____

____

2. How does the size and shape of a plant change as it grows from a seed?

____

____

____

1 The seed sprouts.

2 The seedling has a stem, roots, and leaves.

3 A flower forms.

4 The adult makes new seeds.

Reproducing with Flowers

GO ONLINE Explore the *Life Cycle of an Orange Tree.*

All living things reproduce. To **reproduce** means to make more of an organism's own kind. How a certain organism grows and reproduces is its **life cycle**. A cherry tree makes new seeds in its flowers. Inside the flowers, male parts make pollen, and female parts make eggs. Birds, insects, other animals, and the wind carry pollen from the male parts of one flower to the female parts of another flower. This is called **pollination**.

Cherry trees in an orchard.

Cherries are the fruit that forms around seeds. The fruit protects the seeds as they develop. After a fruit ripens, it may fall to the ground and break open. That releases the seed. Sometimes an animal eats the fruit and deposits the seed in its droppings. When soil and water conditions are right, the seed can germinate. A new cherry tree will grow.

1. Describe the process of pollination.

2. What role does fruit play in a plant's life cycle?

3. What is the next step in the life cycle of a cherry tree after it has grown into an adult?

Reproducing with Cones

A pine tree's life cycle is similar to a flowering plant, but pine trees don't have flowers. Instead, they form seeds inside cones. Pollination takes place when the wind blows pollen from male cones to female cones. Seeds develop in the female cones. When the seeds are ready, the cone opens, and the wind blows the seeds to new places.

Plants that produce cones for reproduction are called conifers. Conifers have needle-shaped leaves. Most conifers keep their needles year-round. Those plants are often called evergreens. However, not all conifers are evergreens. For example, the western larch's needles turn yellow and drop off during autumn.

Pine branches with pollen-producing male cones.

Where do the seeds of pine trees form?

What **patterns** do you notice between the flowering plant and conifer life cycle **models**?

FOLDABLES®

Cut out the Notebook Foldables tabs given to you by your teacher. Glue the anchor tabs as shown below. Use what you have learned and the picture of the pumpkin life cycle to define each word.

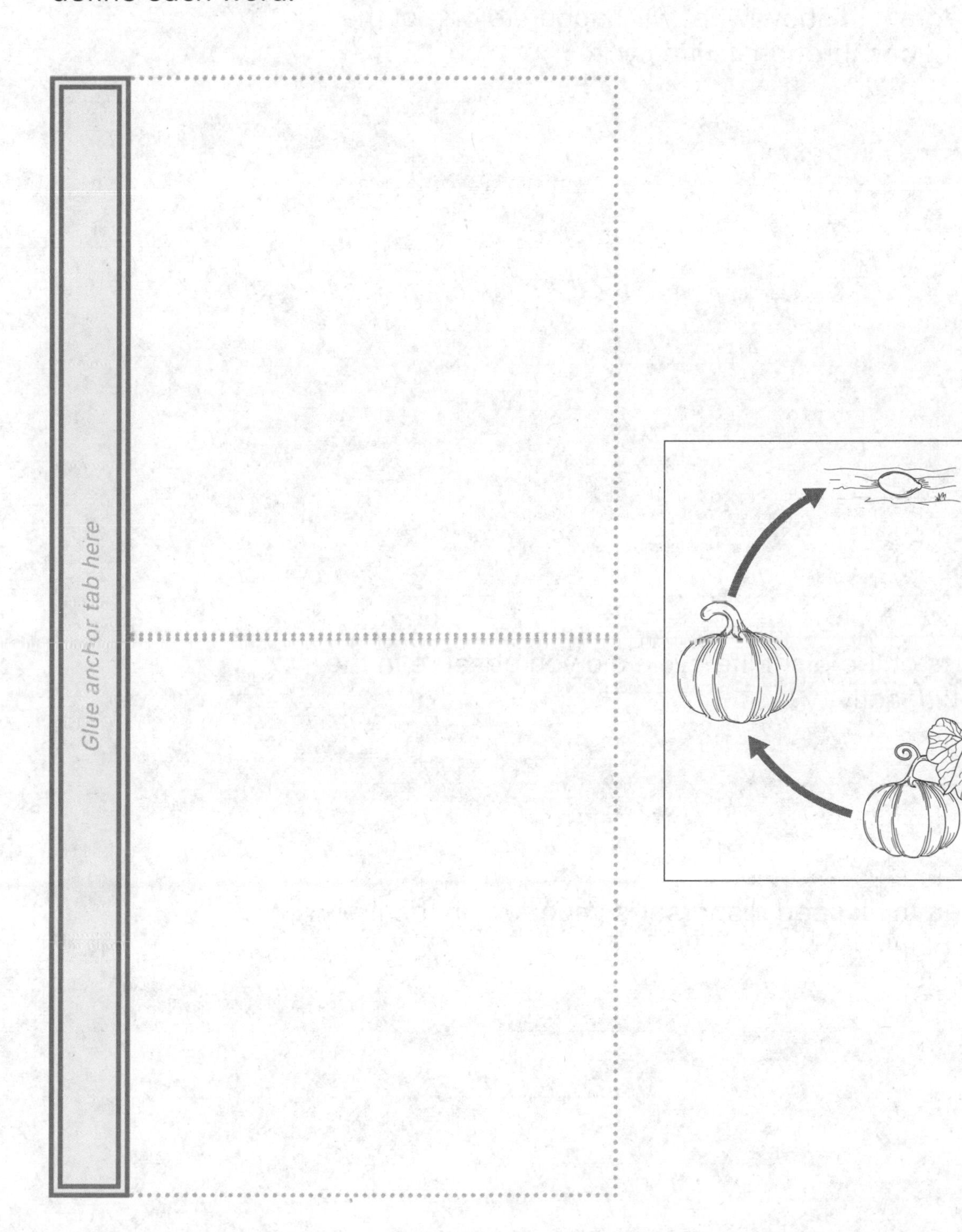

Seed Growth

GO ONLINE Explore the *Seed Dispersal* simulation on how plants spread their seeds.

Revisit the *Seed Growth* activity on pages 8–9. Use what you have learned in the lesson so far to answer the following questions.

1. Draw a diagram to show what will happen to one of the seeds as it goes through its life cycle.

2. Which parts of the plant life cycle did you observe in the *Seed Growth* activity?

3. Why do you think seed dispersal is important in the life cycle of a plant?

Revisit the Page Keeley Science Probe on page 5.

Different Life Cycles

GO ONLINE Watch the video *Strange But True: Plant Life Cycles.*

Have you ever wondered why some plants grow quickly, but others take a long time to grow? For example, you saw that the radish seed germinated in about five days. If you planted it in soil, you would get new radishes in about 21 days. If you planted an apple tree, it would take about six years for the tree to produce new apples. The length of a plant's life cycle depends on its needs and structure.

Plants need sunlight, water, carbon dioxide, nutrients, and space to grow. If a plant gets too little or too much of any of these, it will not grow. Grass plants grow very close together, so they don't need a lot of space. Vegetables grow better if they have room between them.

Weeds do not need a lot of space or nutrients to grow and have very simple structures, so they germinate and grow quickly. Fruit trees have deep roots, leaves, and flowers, so they need a lot of water and sunlight to grow. It may take years for a fruit tree to produce new fruit.

Look at the roots of these two plants. Which plant do you think takes longer to go through its life cycle?

STEM Connection

What Does an Entomologist Do?

Entomologists study insects. They sometimes work with farmers to protect their plants from harmful insects. They focus on getting rid of harmful pests without hurting the beneficial, or helpful, insects. The beneficial insects are those that help with pollination or by eating other, harmful insects.

Bill Hendrix is an entomologist. He has to be aware of plant life cycles. Some insects attack plants when they start to germinate, while others attack the fruit. Some insects will attack a plant at any stage of its life cycle. Bill Hendrix provides farmers with ways of getting rid of the insects at the different stages of the plant's life cycle.

It's Your Turn

Why is it important for an entomologist to work with farmers?

INQUIRY ACTIVITY

Research

Plant Life Cycle Model

You will be preparing a presentation on the life cycle of two plants.

Make a Prediction How will the life cycles of two plants be alike and different?

____

____

____

____

____

Carry Out an Investigation

1. Research the life cycles of two plants.
2. Draw the life cycle models of your two plants.
3. Write about the patterns you see in both life cycles.
4. Write about the differences between the two life cycles.
5. Share your presentation with your class.

Communicate Information

6. **ENVIRONMENTAL Connection** Describe how human activities might influence the life cycle of plants to a classmate.

EXPLAIN THE PHENOMENON

How is the dandelion changing?

Summarize It

Explain the life cycle of flowering plants.

Revisit the Page Keeley Science Probe on page 5. Has your thinking changed? If so, explain how it has changed.

Three-Dimensional Thinking

1. Trees that do not have flowers form seeds in their ______.

A. roots

B. stems

C. leaves

D. cones

2. What stages do all plant life cycles have in common?

3. The stage in which a seed grows into a young plant is called _____.

A. blooming

B. germination

C. reproduction

D. pollination

Extend It

You are a member of the school gardening club. The club has been asked to plan a flower garden that will germinate in the spring and complete a full life cycle in a year. The plants need to grow in an area that receives 4–6 hours of shade. Research information on annuals, biennials, and perennials and decide which type is the best choice for the flower garden. Give examples. Write a speech, draw a poster, create a flyer, or use media to present your information to the class.

OPEN INQUIRY

What questions do you still have?

Plan and carry out an investigation to answer one of the questions.

KEEP PLANNING

STEM Module Project
Science Challenge

Now that you have learned about plant life cycles, go to your Module Project to explain how the information will affect your plan for your garden.

PAGE KEELEY SCIENCE PROBES

LESSON 2 LAUNCH

Flowering Plants

Four friends were looking at the flowering plants in their garden. They wondered how they could tell which plants were most closely related. This is what they said:

Fiona: *I think they should have the same petal shape and flower color.*

Sabrina: *I think they should have the same shaped leaves and petals.*

Adam: *I think they should have the same color of their flowers and leaves.*

Justin: *I think they should have the same height and number of leaves.*

Who do you think has the best idea? ____________________

Explain why you think it is the best idea.

____

____

____

____

You will revisit the Page Keeley Science Probe later in the lesson.

LESSON 2

Plant Traits

ENCOUNTER
THE PHENOMENON

Why is the flower a different color from the rest?

GO ONLINE

Check out *Tulip Field* to see the phenomenon in action.

Talk About It

Look at the photo and watch the video of the tulips. What questions do you have? What observations can you make about the photo? Talk about them with a partner. Record your thoughts below.

Did You Know?

Peaches, pears, strawberries, and apples are all related to roses!

INQUIRY ACTIVITY

Hands On

Plant Families

You observed that different looking plants can be related. Observe different parts of plants and decide if they are related.

State the Claim How can you determine if plants are related?

____

____

Materials

cherry

plum

grape

plate

plastic knife

photo cards

Carry Out an Investigation

BE CAREFUL Use caution when handling the knife.

1. **Record Data** Look at the outside of the cherry, plum, and grape. Record your observations on a separate sheet of paper.
2. Carefully cut each fruit in half on the plate, and observe the seeds. Record your observations.
3. **Record Data** Look at the photos of the leaves and blossoms. Record your observations.

Communicate Information

4. Which two fruits do you think are related? What evidence have you observed that helps with your choice?

____

____

____

____

MAKE YOUR CLAIM

How can you determine if plants are related?

Make your claim. Use your investigation.

CLAIM

I think the plants are related if they ____________________.

Cite evidence from the lesson.

EVIDENCE

I found ____________________.

Discuss your reasoning as a class. Tell about your discussion.

REASONING

The reasoning for my claim is ____________________.

You will revisit your claim later in the lesson.

VOCABULARY

Look for these words as you read:

inherited traits

trait

variation

Traits

Every organism has traits that make it unique. A **trait** is a feature of a living thing. Flower color and plant height are some traits of plants. Traits help you recognize and describe an organism.

An organism's traits show us what kind of living thing it is. We know a cucumber from a rose by its observable traits. Traits are adaptations that help organisms survive. A rosebush has thorns to prevent animals from eating it.

Draw arrows to point to two observable traits of a cucumber and two observable traits of a rose.

Inherited Traits

An apple tree has roots, a trunk, and leaves because its parents did. Heredity is the passing on of traits from parents to offspring. Traits that come from parents are called **inherited traits**. Most living things inherit traits from both parents. An apple tree's ability to make apples is an inherited trait.

GO ONLINE Watch the video *Plant Traits and Survival* to see how variations help plants survive.

Often, parents have different individual traits from each other. Perhaps one plant is red and the other is yellow. The offspring can be red, yellow, or even a different color. An offspring can have traits different from both its parents. Some inherited traits do not show up in parents but do appear in their offspring. These differences are called variations. A **variation** is the appearance of an inherited trait that makes an individual different from other members of the same family.

Where do offspring get their inherited traits?

INQUIRY ACTIVITY

Simulation

Parent Plants

GO ONLINE Explore the *Plant Parents and Offspring* simulation to complete this activity.

Make a Prediction Compare the parent plant to its offspring. Use what you have learned to make a new prediction about how the parent plant and its offspring are alike and different.

Carry Out an Investigation

1. Use the simulation to explore the growth of plants.
2. Complete the data on the table below.

Trait	Parent	Offspring
Height		
Color		
Petal		
Leaves		
Stem		

Communicate Information

3. In the first part of the simulation, you selected two parent plants. How did the offspring compare to the parents?

Talk About It

Did your findings support your prediction? Explain.

WRITING Connection Write an explanatory paragraph about inherited traits. Include information about how variations affect the survival in plants.

COLLECT EVIDENCE

Add evidence to your claim on page 25 about how to determine if plants are related.

Inspect

Read the passage *One Potato, Two Potato, More!* Underline text that tells what was wrong with the first, wild potatoes.

Find Evidence

Reread Highlight evidence that explains how potatoes were changed over time.

Notes

One Potato, Two Potato, More!

In 1535, Spanish soldiers ransacked a deserted village in Peru hoping to find gold. All they found was potatoes. They couldn't have known that this lowly vegetable would be one of the greatest treasures they would send back to Europe.

Wild potatoes were first harvested by South American natives 7,000 years ago. They were small, hard, and bitter. But over time Inca farmers in the Andes Mountains developed larger and tastier potatoes. They planted certain kinds of potatoes that survived well in the cold next to other ones that grew larger or tasted better. These would cross-pollinate to create potatoes that were both hardy and good tasting. By continually experimenting, Inca farmers developed potatoes for every kind of soil and weather condition in the high, windswept mountains. Some of their potatoes

grew well in wet fields; others withstood drought. By the time of the Spanish conquest, the Incas had 3,000 varieties of potatoes to choose from.

Farming has come a long way since the time of the Inca farmer growing red, purple, blue, and yellow varieties of potatoes. Modern farmers change the land to make it suitable for huge crops of the potatoes that have the traits that Americans enjoy.

Make Connections

Talk About It

Look back at the flowers from earlier in the lesson. How might a scientist use this information about variations of potatoes to make variations of flowers?

Notes

REVISIT PAGE KEELEY SCIENCE PROBES Revisit the Page Keeley Science Probe on page 21.

STEM Connection

What Does a Plant Geneticist Do?

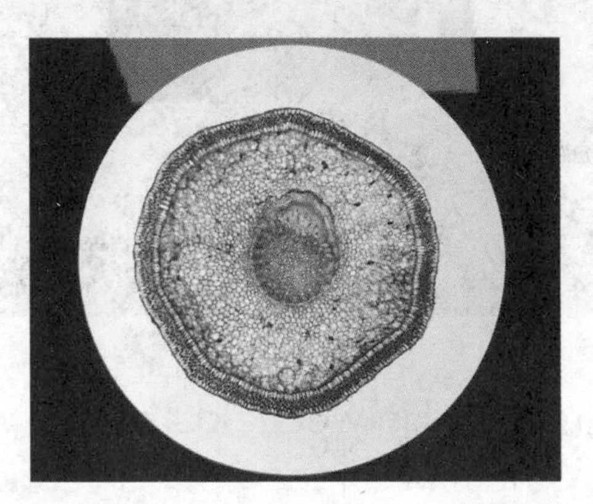

Plant Geneticists do research to find out about new varieties of plants and crops. Geneticists study the traits of plants and animals. Geneticists who study human beings look at a person's DNA. Plant geneticists look at a plant's DNA. By studying the plant's DNA, they can find ways to improve the plant, such as changing what it looks like and how big it grows.

Plant geneticists spend a lot of time working in teams, with members that have many different skills. They write reports and record data. They also work with plants in greenhouses or in the field, and they study how plants are similar to and different from their parents.

It's Your Turn

As a plant geneticist, what could you do to study how a plant compares to its parent?

Pea Plants

When plants and animals reproduce, how can we tell what their offspring will look like? The study of the way living things inherit traits is called genetics. In the 1800s, a man named Gregor Mendel introduced the idea of genetics. He observed that we can often predict the traits offspring will have by looking at the traits of the parents.

Mendel did experiments with pea plants. He predicted what flower color the offspring would likely have based on the flower color of the parents. He found that parents with purple flowers would likely have offspring with purple flowers. He also found that parents with white flowers would likely have offspring with white flowers.

What would happen if plants with purple flowers reproduced with plants with white flowers? At first, all of the offspring had purple flowers. Then, when the new generation of offspring reproduced, the result was a mix. Three out of four offspring had purple flowers. Only one out of four offspring had white flowers. Mendel did thousands of experiments with pea plants to understand how traits could be predicted.

Compare your results of the *Parent Plants* Activity on page 28 to those of Mendel's experiments.

PRIMARY SOURCE

Gregor Mendel experimented with pea plants to understand how traits are passed from parents to offspring.

EXPLAIN
THE PHENOMENON

Why is the flower a different color from the rest?

Summarize It

Explain how parents and their offspring compare.

REVISIT
PAGE KEELEY SCIENCE PROBES

Revisit the Page Keeley Science Probe on page 21. Has your thinking changed? If so, explain how it has changed.

Three-Dimensional Thinking

1. The passing of traits from parents to offspring is called

A. trait

B. inherit

C. heredity

D. variation

2. Does a flower always have the exact same color as its parents? Explain why or why not.

3. Look at the chart below. Which of the following traits belongs in the empty box?

Inherited Traits	
flower color	**stem height**
seed shape	**leaf shape**
	taste of fruit

A. number of seeds

B. type of root

C. number of blooms

D. dry leaves

Extend It

You work in a produce department of a grocery store. You have been asked to give a workshop on apples. You must research how apples developed into what we know today.

KEEP PLANNING

STEM Module Project
Science Challenge

Now that you have learned about plant traits, go to your Module Project to explain how the information will affect your plan for the garden.

STEM Module Project
Science Challenge

Growing Plants

You have been hired as a botanist. Using what you have learned throughout this module, you will choose three crop plants that can be harvested in less than 40 days. Your presentation should include information about the life cycle, traits, growth requirement, and care instruction for each plant.

Planning after Lesson 1

Apply what you have learned about plant life cycles to your project planning.

How does knowing about plant life cycles affect your module planning?

Record information to help you plan your model after each lesson.

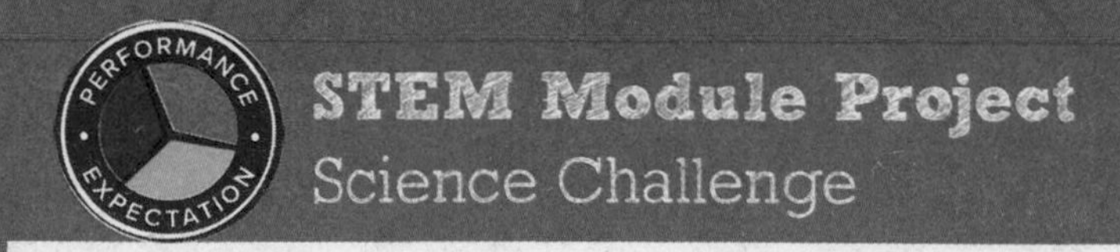

Planning after Lesson 2

Apply what you have learned about plant traits to your project planning.

How can plant traits affect your garden?

Research the Problem

Research information about the life cycle and traits of three crop plants that are easy to grow. Include information about their growth requirements and care instructions by going online to teacher-approved websites, interviewing a botanist, or finding a book about plants from your local library.

Source	Information to Use in My Project

STEM Module Project
Science Challenge

Growing Plants

Look back at the planning you did after each lesson. Use that information to complete your final module project.

Build Your Model

1. Use your project planning to build a model.
2. Write clear steps to design your garden.
3. Determine the materials you will need to design your model and prepare your presentation to the class. List the materials in the space provided.
4. Your plan should meet all the criteria to successfully complete your project.

Materials

Procedure:

Sketch Your Model

Use a model to explain your design solution. Your drawing should include information about the number and type of plants, their growth requirements, and their location in the school garden.

Complete the science challenge!

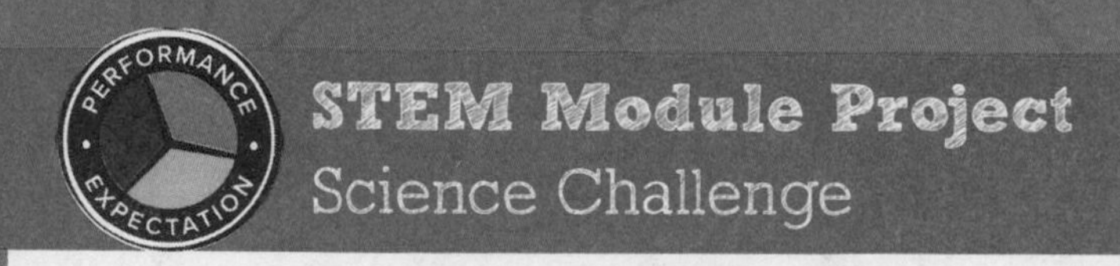

Communicate Your Results

Share the plan for your garden and your findings with another group. Compare the plants and their life cycles. Communicate your findings below.

MODULE WRAP-UP

REVISIT THE PHENOMENON

Use what you learned in this module to explain how plants reproduce.

How have your ideas about plant life cycles changed?

Animals

ENCOUNTER
THE PHENOMENON

How do the goats climb the tree?

GO ONLINE
Check out *Climbing Goats* to see the phenomenon in action.

Talk About It

Look at the photo and watch the video of the goats climbing the tree. What questions do you have? Talk about your observations with a partner.

Did You Know?

The trees can grow up to 32 feet tall. The goats climb all the way to the top because they are attracted to the fruit.

STEM Module Project Launch
Science Challenge

Design a Habitat

You have been hired as a habitat specialist at a zoo. A new animal has just arrived. What will you need to provide for this animal? At the end of the module, you will develop a design for a habitat for the new animal. Your goal will be to design a model that is able to help the animal survive in its new environment.

Lesson 1
Animal Life Cycles

Lesson 2
Animal Traits

Lesson 3
Animal Group Survival

What do you think you need to know before you can design a habitat?

Habitat specialists study animals in the wild so that they can learn what the animals need to survive. They design homes for animals in zoos.

OWEN
Entomologist

STEM Module Project

Plan and Complete the Science Challenge Use what you learn throughout the module to design a habitat.

PAGE KEELEY SCIENCE PROBES

LESSON 1 LAUNCH

Life Cycle Stages

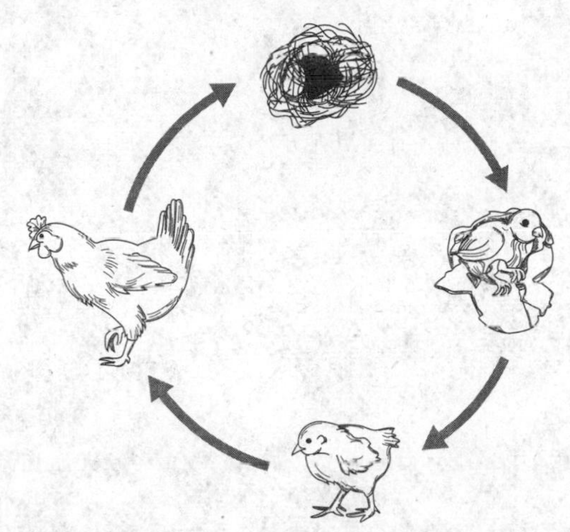

Which animals go through a life cycle similar to the stages of a chicken's life cycle?

Butterfly	Horse	Duck
Snake	Turtle	Lizard
Cat	Crow	Mouse

Explain your thinking. How did you decide if the stages of an animal's life cycle are similar to the stages of a chicken's life cycle?

You will revisit the Page Keeley Science Probe later in the lesson.

LESSON 1

Animal Life Cycles

ENCOUNTER
THE PHENOMENON

Will the cub grow up to look more like the adult mountain lion?

GO ONLINE

Check out *Mountain Lion* to see the phenomenon in action.

Talk About It

Look at the photo and watch the video of the mountain lions. What observations have you made about the the phenomenon? Talk about them with a partner. Record or illustrate your thoughts below.

Did You Know?

The bowhead whale is the longest-living mammal. The oldest known bowhead whale was 211 years old.

INQUIRY ACTIVITY

Hands On

Grow a Caterpillar

You saw a baby and adult mountain lion. Now you will observe how a caterpillar grows and develops into an adult.

Materials

caterpillar kit

hand lens

Make a Prediction How will a caterpillar change as it grows?

____

____

____

Carry Out an Investigation

1. Look at the 5 caterpillars with a hand lens. Draw a picture of them and label all the parts you can see.

2. Record Data Observe the caterpillars for six weeks. Draw and record your observations in the data table. Label any changes you notice throughout the caterpillars' life cycle.

	Observations
Week 1	
Week 2	
Week 3	
Week 4	
Week 5	
Week 6	

Communicate Information

3. What small changes did the caterpillars go through?

4. What big changes did you observe?

Talk About It

Did your findings support your prediction? Did your caterpillar grow as you expected it?

VOCABULARY Look for these words as you read:

birth metamorphosis

Animal Life Cycles

GO ONLINE Watch the video *Metamorphosis* to see a caterpillar change.

The life cycle of every animal includes birth, growth, reproduction, and death.

Some animals change shape through a process called **metamorphosis**. Most insects and amphibians undergo metamorphosis.

Insects

Insects start life as eggs. The egg contains food that the young animal needs. When the young animal has grown enough, it hatches, or breaks out of the egg and becomes larvae. Next, the larvae turn into pupa. The pupa stage is when the most changes take place. Because there are so many changes happening to the insect's body, it does not move much during the pupa stage. Eventually, it grows into an adult that can have its own young. Use your information from the *Grow a Caterpillar* activity to draw the stages of a caterpillar's life cycle. Label each stage using the terms *egg, larva, pupa,* and *adult*.

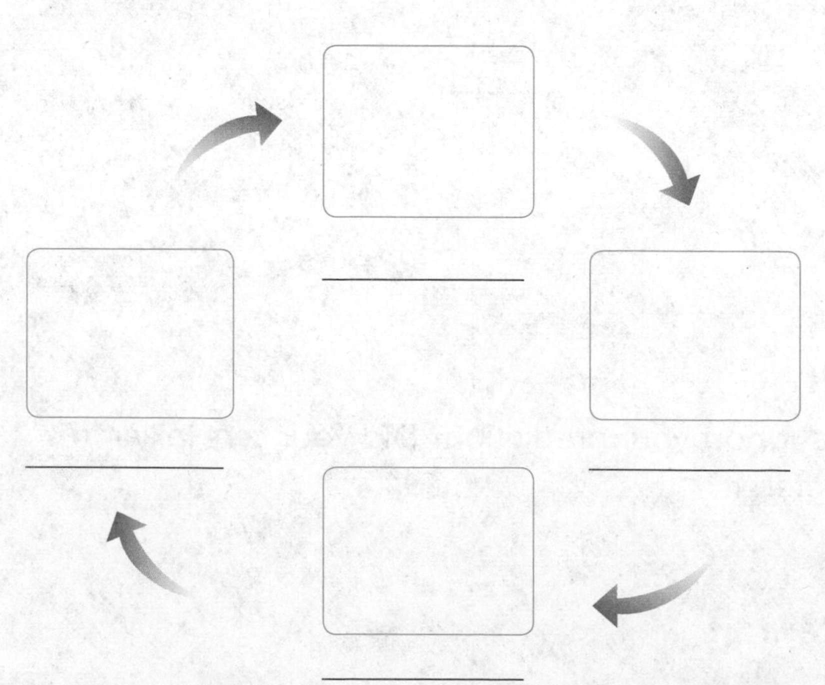

Amphibians

The frog is an amphibian that undergoes metamorphosis. Amphibians are animals that live part of their lives in water and part on land. Look at the photos of a frog's life cycle. Follow the arrows and describe the frog's life cycles.

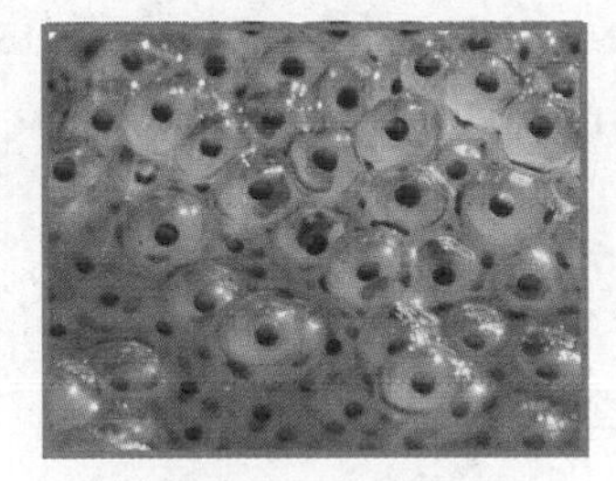

Egg Frogs lay eggs in water.

Tadpole Young frogs, or tadpoles, hatch. Like fish, they swim and breathe with gills.

Becoming an Adult A tadpole starts to grow legs and lungs.

Adult Now the frog looks like its parents. It moves onto land and can reproduce.

What is a similarity and a difference between the butterfly's and frog's life cycles?

INQUIRY ACTIVITY

Hands On

Mealworms

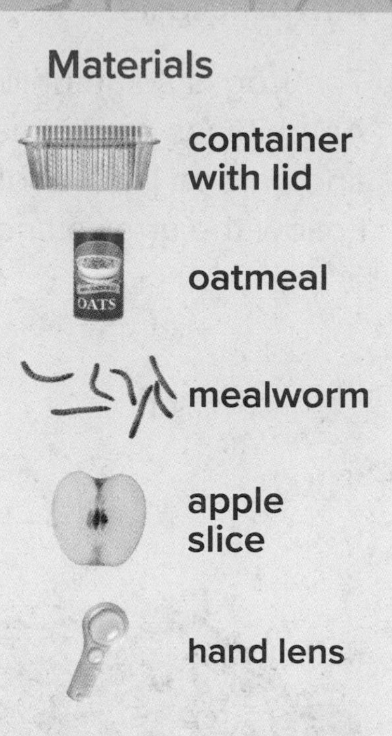

You have learned about the life cycles of various animals. Mealworms hatch from eggs and then grow into adults.

Make a Prediction How does a mealworm change as it grows?

____

____

____

Carry Out an Investigation

BE CAREFUL Handle animals with care.

1. Put some oatmeal in the container with the lid.
2. Gently place a mealworm and an apple slice in the container.
3. **Record Data** Observe the mealworm with a hand lens two or three times a week. Record your observations once a week for eight weeks.

Week	Color	Segments	Drawing
1			
2			
3			
4			

week	Color	Segments	Drawing
5			
6			
7			
8			

Communicate Information

4. Draw a picture of the life cycle of a mealworm.

5. Did your findings support your prediction? How did the mealworm change shape as it grew?

Talk About It

What type of life cycle do mealworms undergo? Explain.

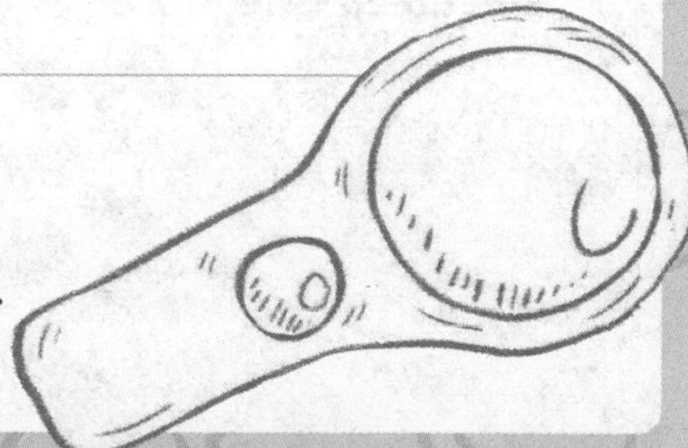

More Animal Life Cycles

Reptiles, Fish, and Birds

> **GO ONLINE** Watch the video *Strange Animal Life Cycles* to see the life cycles of various animals.

When many animals begin their life, they look like smaller versions of their parents. Most reptiles, fish, and birds lay eggs. When these eggs hatch, the babies look like small versions of their parents. This means that they do not undergo metamorphosis. Reptiles, fish, and birds have similar life cycles, but there are some differences. Birds and reptiles lay eggs on land, and fish lay eggs in water. Most birds look after their young, but most fish and reptiles do not.

A robin starts its life as an egg, hatches, and becomes an adult bird.

1. How do the life cycles of most birds, fish, and reptiles begin? What other animal groups also begin life this way?

____

____

____

2. What are some differences in life cycles of birds, reptiles, and fish?

____

____

____

Mammals

Like reptiles, fish, and birds, young mammals look a lot like their parents. This similarity means they do not undergo metamorphosis. Most mammals give birth to live young. **Birth** is the beginning or origin of a plant or animal.

Many young mammals are weak and slow. They cannot find their own food. They need their parents in order to survive. Mammal parents care for their young in different ways. For example, kangaroos protect and carry their young in their pouch for up to eleven months. Parent beavers build strong homes to keep their young safe from predators.

What patterns did you observe between the **life cycles** of mammals and the **life cycles** of birds, fish, and reptiles?

REVISIT PAGE KEELEY SCIENCE PROBES Revisit the Page Keeley Science Probe on page 47.

STEM Connection

What Does a Wildlife Biologist Do?

Wildlife Biologist study animals and how they interact with their environment. They spend a lot of time in field doing research on large animals, many animals need to be sedated to ensure saftey of the biologist and animal. Wildlife biologists are good with animals, and they research how animals grow and develop.

Wildlife biologists do a lot of calculations and create a lot of charts and graphs. For example, they do calculations to figure out what percentage of an animal's diet comes from fruit and nuts..

It's Your Turn

As a wildlife biologist, what information or data would you need to have in order to create the right habitat for an animal?

INQUIRY ACTIVITY

Research

Animal Life Cycle Model

Prepare a presentation on the life cycle of two different animals. Pick two different animal groups from this list: insect, fish, bird, or mammal. Then identify one animal from the groups you selected for your research.

Make a Prediction What patterns will you observe in the life cycles of the animals you selected?

Carry Out an Investigation

1. Research the life cycles of your chosen animals.
2. Draw the life cycle models of your two animals.
3. Explain how the life cycles of your two animals are alike and different.
4. Did what you learned support your prediction? Explain.

Communicate Information

5. Share your presentations with the class.

ENVIRONMENTAL Connection

6. On a separate sheet of paper explain how might human activities change the cycle of the animals?

EXPLAIN THE PHENOMENON

Will the cub grow up to look more like the adult mountain lion?

Summarize It

Explain how different animals grow and develop.

Revisit the Page Keeley Science Probe on page 47. Has your thinking changed? If so, explain how it has changed.

Three-Dimensional Thinking

1. What stages do all animal life cycles have in common?

2. A pattern is something that repeats. Do you think a life cycle is a pattern? Explain.

Extend It

Using resources provided by your teacher, conduct research on the importance of butterflies. Monarch butterfly populations are decreasing. Explain what can be done to increase their population. How might a school garden help? Create a plan for a school garden.

KEEP PLANNING

STEM Module Project
Science Challenge

Now that you have learned about animal life cycles, go to your Module Project to explain how the information will affect your plan for your habitat.

PAGE KEELEY SCIENCE PROBES

LESSON 2 LAUNCH

Sadie's Poodle

Sadie has a large, white poodle. Her poodle does tricks, barks at strangers, and wags her tail when she is happy. Sadie and her friends have different ideas about the poodle's size, her fur color, and her behaviors. Here is what they said:

Sadie: *My poodle gets all of her characteristics from her parents.*

Josh: *I think your poodle gets some of her characteristics from her parents and some from interacting with her environment.*

Omar: *I think your poodle gets all of her characteristics from interacting with her environment.*

Who do you agree with the most? ______________________

Explain why you agree.

____

____

____

You will revisit the Page Keeley Science Probe later in the lesson.

LESSON 2

Animal Traits

ENCOUNTER
THE PHENOMENON

Why do the kittens look different from the mom and each other?

GO ONLINE

Check out *Cat Litter* to see the phenomenon in action.

Talk About It

Look at the photo and watch the video of the mom cat and her litter. What observations and questions do you have about the phenomenon? Talk about them with a partner. Record your thoughts below.

Did You Know?

No two tigers ever have the same stripes.

INQUIRY ACTIVITY

Data Analysis

Inherited Traits

You observed a litter of cats that looked different than their mother. Dogs and other animals inherit traits from their parents too. Analyze the traits passed down to dogs from their parents by interpreting the photos in the activity.

Make a Prediction Which observable traits were passed down from the parent dogs to their offspring?

Carry Out an Investigation

1. Examine the observable traits of the parent dogs.
2. **Record Data** Record your observations in the data table.
3. Examine the observable traits of the offspring.
4. **Record Data** Record your observations in the data table.

poodle (parent)

yellow labrador (parent)

labradoodle (offspring)

brown labradoodle (offspring)

	Observable Traits
Poodle (parent)	
Yellow Labrador (parent)	
Labradoodle (offspring)	
Brown Labradoodle (offspring)	

Communicate Information

5. Look at the data you collected. What traits did the labradoodle inherit from its parents?

6. What traits did the brown labradoodle inherit from its parents?

Talk About It

Neither the poodle nor the labrador has brown fur. Discuss with a classmate why you think one of the labradoodles has brown fur.

Animal Reproduction

VOCABULARY

Look for these words as you read:

environmental trait

instinct

learned trait

There are two ways animals reproduce. Some animals reproduce by budding. A bud forms on the adult's body. After the bud breaks off, it grows into an adult. It is a copy of its parent. The offspring has exactly the same characteristics, or traits, as the parent.

Another kind of reproduction requires cells from both a male parent and a female parent. When the female and male cells join, offspring are produced. This new offspring has traits from both its parents. It is not identical to either parent.

Why do you think the offspring of budding are identical to the parent, but the offspring of two parents are not?

____

____

____

____

____

Like all mammals, giraffes reproduce with two parents.

Inherited Traits

GO ONLINE Explore *Compare Caterpillars* to compare the traits of two caterpillars.

You have learned about traits of plants. Animal traits work similarly. Animals inherit their traits from their parents. Your eye color and hair color are inherited traits. The number of legs an animal has is also inherited. Inherited traits make animals look like their parents.

poodle

Inherited traits can also affect the way an animal acts. Humans can inherit color-blindness, a trait that is not visible. Color blindness is when you cannot see certain colors correctly. Reflexes, such as blinking, are actions that parents pass on to their offspring. So are instincts inherited behaviors? An **instinct** is a way of acting that an animal does not have to learn. Birds build nests and spiders spin webs by instinct. Many instincts help animals survive. Inherited behavior is a set of actions that parents pass on to their offspring. Birds build nests in which they lay their eggs and raise their young.

labradoodle

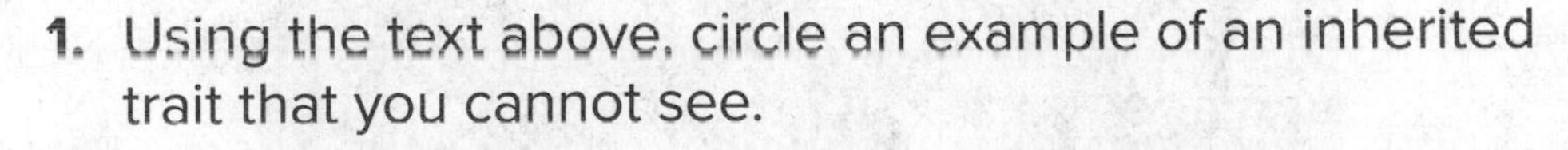

1. Using the text above, circle an example of an inherited trait that you cannot see.

Look at the photos of the dogs. The labradoodles have one poodle parent and one yellow Labrador parent.

yellow labrador

Think back to the *Inherited Traits* activity. What **patterns** do you notice when **analyzing** the **inherited traits** of the dog offspring?

brown labradoodle

Inspect

Read the passage *Inherited Physical and Behavioral Traits*. Underline the text that describes the physical and behavioral traits of the galagos.

Find Evidence

Reread Using the text you underlined, sort the traits by physical and behavioral traits.

Notes

Inherited Physical and Behavioral Traits

You have learned that offspring can inherit traits. In this passage, you will read about physical and behavioral traits. A physical trait is a feature of one's body. Look at the photo on the page. What physical traits can you see?

Behavioral traits are quite different. These traits describe the way one behaves, or acts. Let's read more about the animal in the photo, the galago. The galago is a primate that lives on the continent of Africa.

Galagos are nocturnal. They are awake at night. Galagos have large eyes to help see in the dark. They have large ears that give them exceptional hearing. They have large feet and toes to grip branches and catch small animals for food. Galagos spend hours grooming and playing in small groups. They search for food alone. They have long tails to help balance while climbing trees. While searching for food, galagos move quickly through trees both on four legs and on two.

Describe a way in which a physical trait and a behavioral trait work together to help galagos survive.

Make Connections

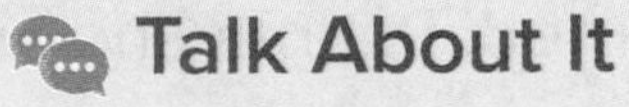

Talk About It

How do these traits help galagos stay safe from predators?

Notes

Learned and Environmental Traits

GO ONLINE Check out *Animal Traits and Survival* to see how inherited, learned, and environmental traits help animals.

Some of your traits come from your parents; others you have learned. People and animals can gain new skills over time. These new skills are called **learned traits**. Learning to ride a bicycle or play the piano are learned traits.

Some traits are affected by the environment. For example, a rabbit's fur may grow thick during the cold winter months. The rabbit may shed its thick fur when the temperature warms in the summer. These types of traits are **environmental traits**.

Learned traits and environmental traits are not passed from parents to offspring. Your parents may know how to ride a bike, but you still had to learn that skill yourself. If your parent has a scar, you were not born with the same scar.

Both the riders and the horses are demonstrating learned traits. The people learned how to ride horses. The horses have learned how to be ridden.

Woodpecker finch using a stick as a tool.

1. How is a learned trait different from an inherited trait?

2. Fur color and type is an inherited trait. How does a polar bear's fur help it survive?

3. Would it be beneficial for a coyote living in a desert to have the same fur as a polar bear? Why?

Revisit the Page Keeley Science Probe on page 63.

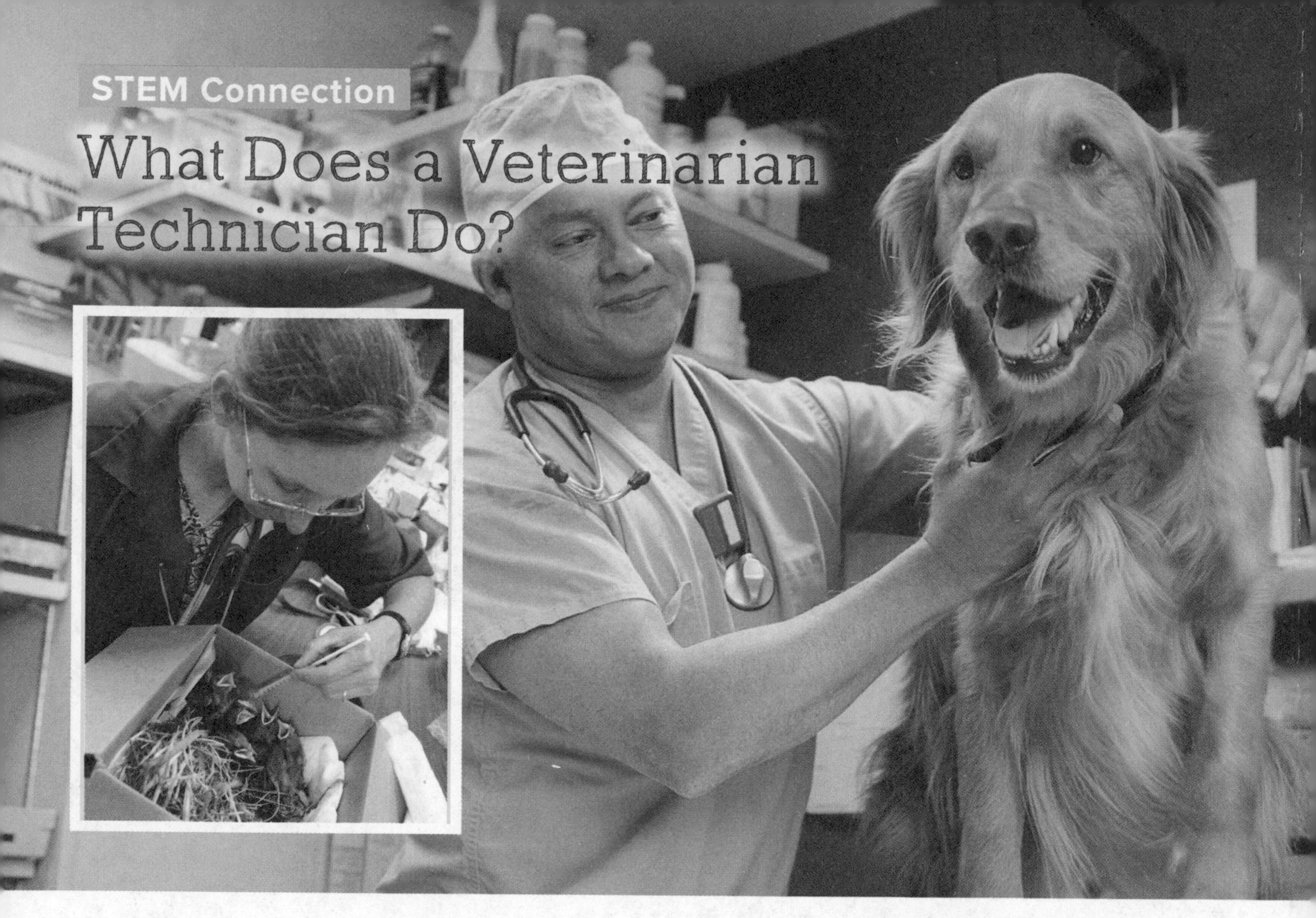

STEM Connection

What Does a Veterinarian Technician Do?

Veterinarian Technicians help doctors who take care of animals. They help the veterinarian with diagnosing and treating animals. Some technicians perform tests and x-rays on the animals that are sick. Some technicians prepare animals for surgery and assist the veterinarian during surgery.

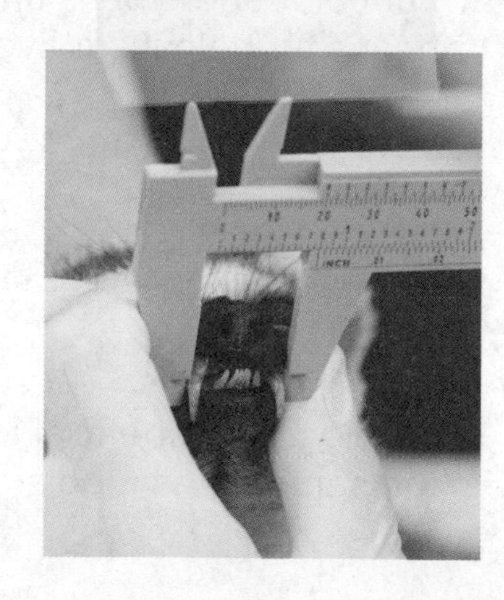

There are some veterinarian technicians who are especially interested in studying how animals compare to their parents. You have probably heard of different breeds of dogs. Animal breeding is the practice of choosing animals to be parents so that there will be a particular trait in the offspring.

It's Your Turn

What similarities have you noticed between an animal and its parents or siblings? What are some traits or characteristics of one type of animal?

INQUIRY ACTIVITY

Simulation

Traits and Survival

GO ONLINE

Use the simulation *Choose Traits for Survival* to determine how inherited, learned, and behavioral traits help a bird survive.

State the Claim What traits help a bird survive?

Carry Out an Investigation

1. Read the description of the environment.
2. Find a trait that will help the bird find food.
3. Find a trait that will help the bird find a mate.
4. Find a trait that will help the bird protect its nest.

Communicate Information

5. Did your results support your claim? Explain.

Talk About It

Describe the traits that helped the bird survive. Did some traits help it more than others?

EXPLAIN
THE PHENOMENON

Why do the kittens look different from the mom and each other?

Summarize It

Explain how animals are similar and different from their parents.

Revisit the Page Keeley Science Probe on page 63. Has your thinking changed? If so, explain how it has changed.

Three-Dimensional Thinking

1. What is another name for an organism's young?

   A. egg

   B. seedling

   C. offspring

   D. pupa

2. What is an example of a learned trait?

   A. a dog running

   B. a bird eating worms

   C. a bear hibernating

   D. a person using a hammer

3. The features that are passed from a parent to offspring are called __________.

   A. environmental traits

   B. inherited traits

   C. learned traits

   D. offspring traits

Extend It

Your teacher has asked several small groups in your class to create a board game. Create a game that includes cards and your knowledge of traits. You may use digital resources provided by your teacher and recyclable materials to build a model.

OPEN INQUIRY

What questions do you still have about traits?

Plan and carry out an investigation to answer one of the questions.

KEEP PLANNING

STEM Module Project
Science Challenge

Now that you have learned about animal traits, go to your Module Project to explain how the information will affect your plan for your habitat.

PAGE KEELEY SCIENCE PROBES

LESSON 3 LAUNCH

Animal Groups

Many animals interact with each other in groups. Put an X on all the boxes that describe how animals interact in groups.

Animals interact in groups when they need to find food.	**Animals interact in groups when they need to defend themselves.**
Animals interact in groups when they need to cope with changes.	**Animals sometimes interact in small groups.**
Animals sometimes interact in large groups.	**Animals only interact in pairs.**
Animals only interact with animals of their own kind.	**Animals sometimes interact with other kinds of animals to survive.**

Explain your thinking. Describe your ideas about how animals interact in groups.

You will revisit the Page Keeley Science Probe later in the lesson.

LESSON 3

Animal Group Survival

ENCOUNTER
THE PHENOMENON

Why are the fish swimming in a circle?

GO ONLINE

Check out *School of Fish* to see the phenomenon in action.

Talk About It

Look at the photo and watch the video of the school of fish. What observations and questions do you have about the phenomenon? Talk about them with a partner. Record or illustrate your thoughts below.

Did You Know?

Some fish live in colonies, in large groups, just like ants. One type of ant can have 700,000 ants in its colony!

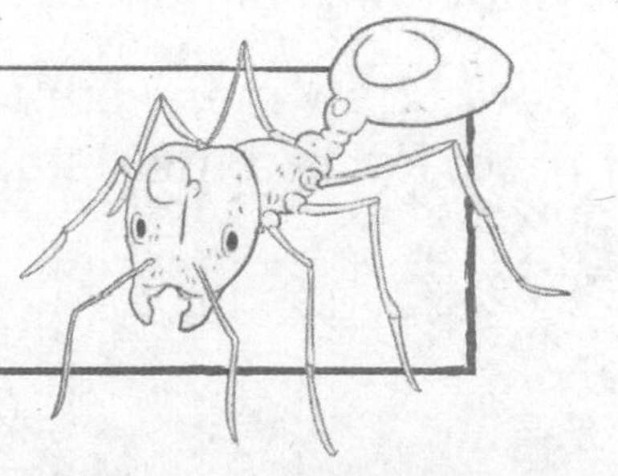

INQUIRY ACTIVITY

Hands On

Ant Workers

Materials

craft sticks

stopwatch

You saw fish working together to hunt for food. Ants also work together to collect food, build structures, and protect their nests. Build a road on your own and then in a group, and compare the results.

Make a Prediction Will you be able to build a longer road by yourself or as a group? Explain.

Carry Out an Investigation

1. Place the craft sticks in a pile on one side of the room. Stand on the other side of the room.
2. You will be timed for one minute. When the timer starts, walk to the pile of craft sticks, and bring back one stick. Place it on your desk to begin building a road across it. Repeat as many times as possible until time is up. Bring back only one stick at a time.
3. **Record Data** Count how many craft sticks were used, and describe your road. Record the data in the table on the next page.

Number of Students	Number of Craft Sticks in the Road	Description of How the Road was Built
1		
4		
4		

4. Repeat steps 1–3, but this time have four students gather craft sticks and work on the road at the same time.

5. Repeat step 4, but this time try to work together in a different way.

INQUIRY ACTIVITY

Communicate Information

6. Did your findings support your prediction? How did the road that you built by yourself compare to the road that you built as part of a group?

7. What benefits would there be for ants in a colony to work together? Use your data to construct an explanation.

8. What different way did you try to work as a team?

MAKE YOUR CLAIM

Why do some animals live in groups?

Make your claim. Use your investigation.

CLAIM

I think the animals work in groups ______________________.

Cite evidence from the lesson.

EVIDENCE

I found that ______________________________________ .

Discuss your reasoning as a class. Tell about your discussion.

REASONING

The evidence supports the claim because ______________________.

 You will revisit your claim later in the lesson.

VOCABULARY Look for these words as you read:

group population survive

Animal Groups

Some animals **survive**, or stay alive, better when they work together as a **group**. A group is a number of living things that have some type of natural relationship. Hyenas work as a pack to hunt their prey. Their large numbers can scare lions away. Each pack member helps raise the hyena cubs. Working together helps the population of hyenas grow and survive. A **population** is all the members of one type of organism that are in the same environment.

GO ONLINE Explore *Ant Colony* to discover how ants live in groups.

In the ocean, certain types of fish will work together to avoid being eaten by a predator. Schools of sardine fish swim together to confuse marlin fish that are trying to catch them.

1. What are some advantages of living in a group?

2. What do you think would happen if an animal became separated from its group?

Flying Solo

GO ONLINE Watch the video *Animal Groups* on the way different animals work together in groups.

While many animals are social and work well in groups, there are some animals that are more likely to survive by living alone. A tiger is a large predator that needs to eat a lot of meat for survival. A tiger can bring down large prey on its own, but it requires a lot of energy. A tiger will need to eat most of the prey to get that energy back. Hunting in a group would mean the tiger would have to share its meal. It is much easier for tigers to hunt alone. Being part of a group would not help a tiger survive.

Some animals, like skunks and snakes, live alone most of the year. They are able to hunt and defend themselves without the help of others. When the weather gets cold in the winter months, these animals get together in large groups for warmth. Although they live alone most of the year, these animals are able to take advantage of group survival when their environment is most harsh.

1. What are the advantages of hunting alone instead of hunting in a group?

2. Choose two groups of animals from the video. How are these two groups the same? How are they different?

All Different Sizes

The size of the group that an animal needs to survive can vary greatly. Some animals work in groups with just a few other animals of their kind. Other animals work together in huge numbers. The size of the group usually depends on the group's needs.

A wolf pack needs to be big enough to work together to bring down large prey. Once the prey is caught, the whole group must share it. Each wolf needs to get enough to eat, so the pack cannot be too large. The pack will keep a certain number of wolves to meet those needs. Most packs have an average of 6–10 individuals, although some may include as many as thirty wolves.

Some birds form flocks of hundreds of individual birds. The more birds there are, the less likely it is that one bird will get eaten or killed by a predator. The large group can also confuse a predator and make it harder to select one bird to go after. The larger the group is, the better it is for each member of the group. For these birds, the need to avoid predators in order to survive means that a very big group is best.

COLLECT EVIDENCE

Add evidence to your claim on page 85 about why some animals live in groups.

FOLDABLES®

Cut out the Notebook Foldables tabs given to you by your teacher. Glue the anchor tabs as shown below. Use what you have learned to describe and compare these two animal groups.

Glue anchor tab here.

Glue anchor tab here.

INQUIRY ACTIVITY

Hands On

Zebrafish Observations

You have learned about the ways animals benefit from living in groups. You can see group behavior in action by studying zebrafish.

Make a Prediction What do you think will happen if you introduce a new zebrafish into a school of zebrafish?

____

____

____

____

Materials

aquarium kit with zebrafish

water

cup

fish food

Carry Out an Investigation

BE CAREFUL Always handle live animals with care. Wash your hands before and after handling.

1. Scoop some water and one zebrafish out of the aquarium your teacher has set up.
2. Observe the behavior of the zebrafish.
3. Slowly pour the zebrafish from the cup into the aquarium.
4. Observe the behavior of the added zebrafish.
5. Add some fish food to the aquarium. Observe the behavior of the fish.
6. Wearing a glove, place your hand in the water near the zebrafish. Observe how the zebrafish react to your hand.

Communicate Information

7. Describe the behavior of the school of zebrafish.

8. What happened when you added the new zebrafish to the container? Did the results support your prediction?

9. Describe the behavior of the zebrafish when food was added to the container.

10. What happened when you placed your hand in the container? What might cause the zebrafish to react this way in the wild?

STEM Connection

What Does an Animal Behaviorist Do?

Animal Behaviorists work with animals to find out why animals behave the way they do. They spend a lot of time studying animals in their natural environment and observing their behavior. They look for causes of specific behaviors, and they find ways to change behaviors that are unhealthy.

Dr. Cathleen Cox is the head of research at a zoo in California. She uses her knowledge of animal groups to know which animals benefit from living together and which animals need to live alone. For example, the zoo houses a group of 18 male and female chimpanzees that includes great-grandmothers, grandmothers, mothers, fathers, aunts, uncles, brothers, and sisters that all live together as they would in the wild.

It's Your Turn

If you were designing a habitat for a group of zoo animals, which animals would you put in the same environment? What makes you think these animals would survive if they shared the same space?

INQUIRY ACTIVITY

Research

Humpback Whale

Humpback whales can be found off the California coast. Research how humpback whales live, eat, and migrate.

Ask a Question Ask a question you would like your research to answer.

Carry Out an Investigation

1. Use the resources provided by your teacher to research an answer to your question.

Communicate Information

2. When do humpback whales interact with other humpback whales?

Newborn humpback calves are only about the size of the mother's head.

Talk About It

Talk to a partner about your results.

EXPLAIN THE PHENOMENON

Why are the fish swimming in a circle?

Summarize It

Explain why some animals live in groups and others live alone.

REVISIT Revisit the Page Keeley Science Probe on page 79. Has your thinking changed? If so, explain how it has changed.

Three-Dimensional Thinking

1. What are some reasons animals live in groups? Circle all that apply.

A. To hunt for food together

B. To care for their young

C. To not be lonely

D. To protect themselves from predators

2. Why do some animals live alone?

____

____

____

____

3. **WRITING Connection** Explain why some animals live in large groups while others live in small groups.

____

____

____

____

____

____

____

____

____

____

Extend It

You have been asked to make a poster for a local zoo. Research two animals and explain how each animal survives in their environment. The poster will be used at a presentation to preschool students.

KEEP PLANNING

STEM Module Project
Science Challenge

Now that you have learned about animal group survival, go to your Module Project to explain how the information will affect your plan for your habitat.

STEM Module Project

Science Challenge

Design a Habitat

You have been hired as a habitat specialist at a zoo. Using what you have learned throughout this module, you will help the zoo design a habitat for a new animal at the zoo.

Planning after Lesson 1

Apply what you have learned about animal life cycles to your project planning.

How does knowing about animal life cycles affect your habitat design?

Record information to help you plan your model after each lesson.

Planning after Lesson 2

Apply what you have learned about animal traits to your project planning.

How will the traits of the animal affect your design? Give an example.

Planning after Lesson 3

Apply what you have learned about animal group survival to your project planning.

Why do you have to consider the animal's group survival needs?

Research the Problem

Choose an animal. Then gather information about its life cycle, basic needs for growth, group size, and inherited traits by using teacher-approved websites, or finding books at your local library. Be sure to find information on habitat components, such as: space, shelter, and animal enrichment programs.

Source	Information to Use in My Project

PERFORMANCE EXPECTATION

STEM Module Project

Science Challenge

Design a Habitat

Look back at the planning you did after each lesson. Use that information to complete your final module project. Your habitat must be the right size for your animal and must meet the animal's needs at every stage in its life cycle. Your habitat must have the right temperature, shelter, water, and food for your chosen animal. Your habitat must include as many animals as are needed to help your animal survive.

Build Your Model

1. Research your animal's life cycle and behaviors.
2. Use your research to design a habitat for your chosen animal. Your habitat should meet the needs of the animal as it grows and develops. Be sure to consider the traits and behaviors of your chosen animal.
3. List the materials you will need to build a model of your design on the space provided.
4. Build your model using your design. Add labels to your model to point out how the habitat meets the needs of your chosen animal.

Materials

____

____

____

____

____

____

____

____

____

____

Procedure:

____

____

____

____

Design Your Solution

Draw your habitat design. Label the features of your design.

Communicate Your Results

Share the plan for your model and your results with another group. Compare your habitats. Communicate your findings below.

MODULE WRAP-UP

REVISIT THE PHENOMENON

Using what you learned in this module, explain animal traits.

Have your ideas changed? Explain.

Science Glossary

A

adaptation a structure or behavior that helps an organism survive in its environment

atmosphere a blanket of gases and tiny bits of dust that surround Earth

attract to pull toward

axis an imaginary line through Earth from the North Pole to the South Pole

B

balanced forces forces that cancel each other out when acting together on an object

birth the beginning or origin of a plant or animal

C

camouflage an adaptation that allows an organism to blend into its environment

climate the pattern of weather at a certain place over a long period of time

competition the struggle among organisms for water, food, or other resources

D

direction the path on which something is moving

distance how far one object or place is from another

E

ecosystem the living and nonliving things that interact in an environment

electrical charge the property of matter that causes electricity

environmental trait a trait that is affected by the environment

extinction the death of all of one type of living thing

F

floodwall a wall built to reduce or prevent flooding in an area

force a push or pull

fossil the trace of remains of living thing that died long ago

friction a force between two moving objects that slows them down

G

germinate to begin to grow from a seed to a young plant

group a number of living things having some natural relationship

H

hibernation to rest or go into a deep sleep through the cold winter

I

inherited trait a trait that can be passed from parents to offspring

instinct a way of acting that an animal does not have to learn

invasive species an organism that is introduced into a new ecosystem

L

learned trait a new skill gained over time

levee a wall built along the sides of rivers and other bodies of water to prevent them from overflowing

life cycle how a certain kind of organism grows and reproduces

lightning rod a metal bar that safely directs lightning into the ground

M

magnet an object that can attract objects made of iron, cobalt, steel, and nickel

magnetic field the area around a magnet where its force can attract or repel

magnetism the ability of an object to push or pull on another object that has the magnetic property

metamorphosis the process in which an animal changes shape

migrate to move from one place to another

mimicry an adaptation in which one kind of organism looks like another kind in color and shape

motion a change in an object's position

N

natural hazard a natural event such as a flood, earthquake, or hurricane that causes great damage

P

pole one of two ends of a magnet where the magnetic force is strongest

pollination the transfer of pollen from the male parts of one flower to the female parts of another flower

population all the members of a group of one type of organism in the same place

position the location of an object

precipitation water that falls to the ground from the atmosphere

R

repel to push away

reproduce to make more of their own kind

resource a material or object that a living thing uses to survive

S

season one of the four parts of the year with different weather patterns

static electricity the build up of an electrical charge on a material

survive to stay alive

speed a measure of how fast or slow an object moves

T

temperature a measure of how hot or cold something is

trait a feature of a living thing

U

unbalanced forces forces that do not cancel each other out and that cause an object to change its motion

V

variation an inherited trait that makes an individual different from other members of the same family

W

weather what the air is like at a certain time and place

Index

R

S

T

V

W

Z

Dinah Zike's Visual Kinesthetic Vocabulary®

cut on all dashed lines — fold on all solid lines

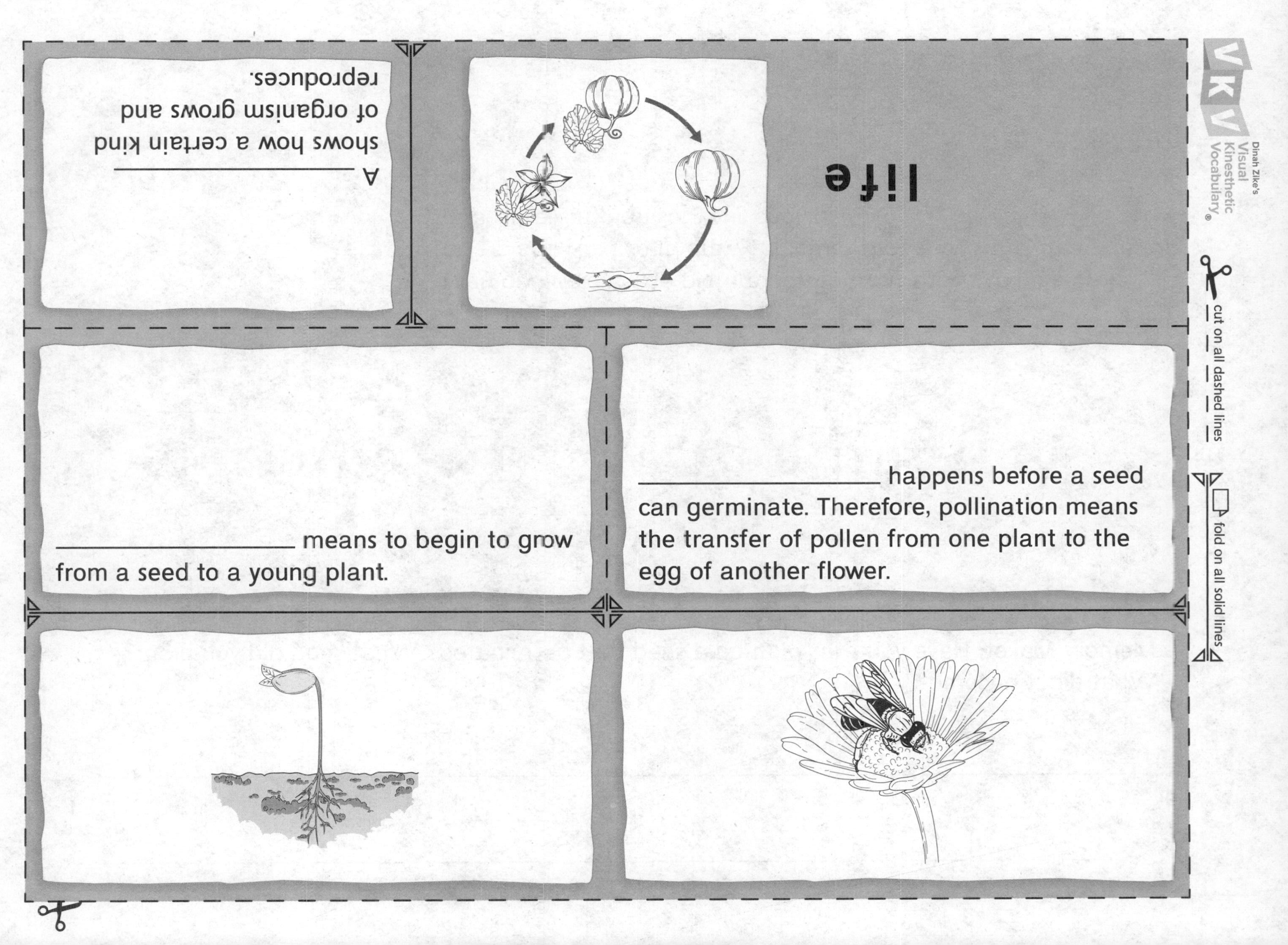

cut on all dashed lines

fold on all solid lines

Memory Maker: Have you ever planted a seed that germinated? What seed did you plant? What did it become?

pollination

germinate

Memory Maker: The picture on the front shows the **life cycle** of a plant. Next to it, draw a picture that shows the life cycle of a different plant of your choice.

cycle

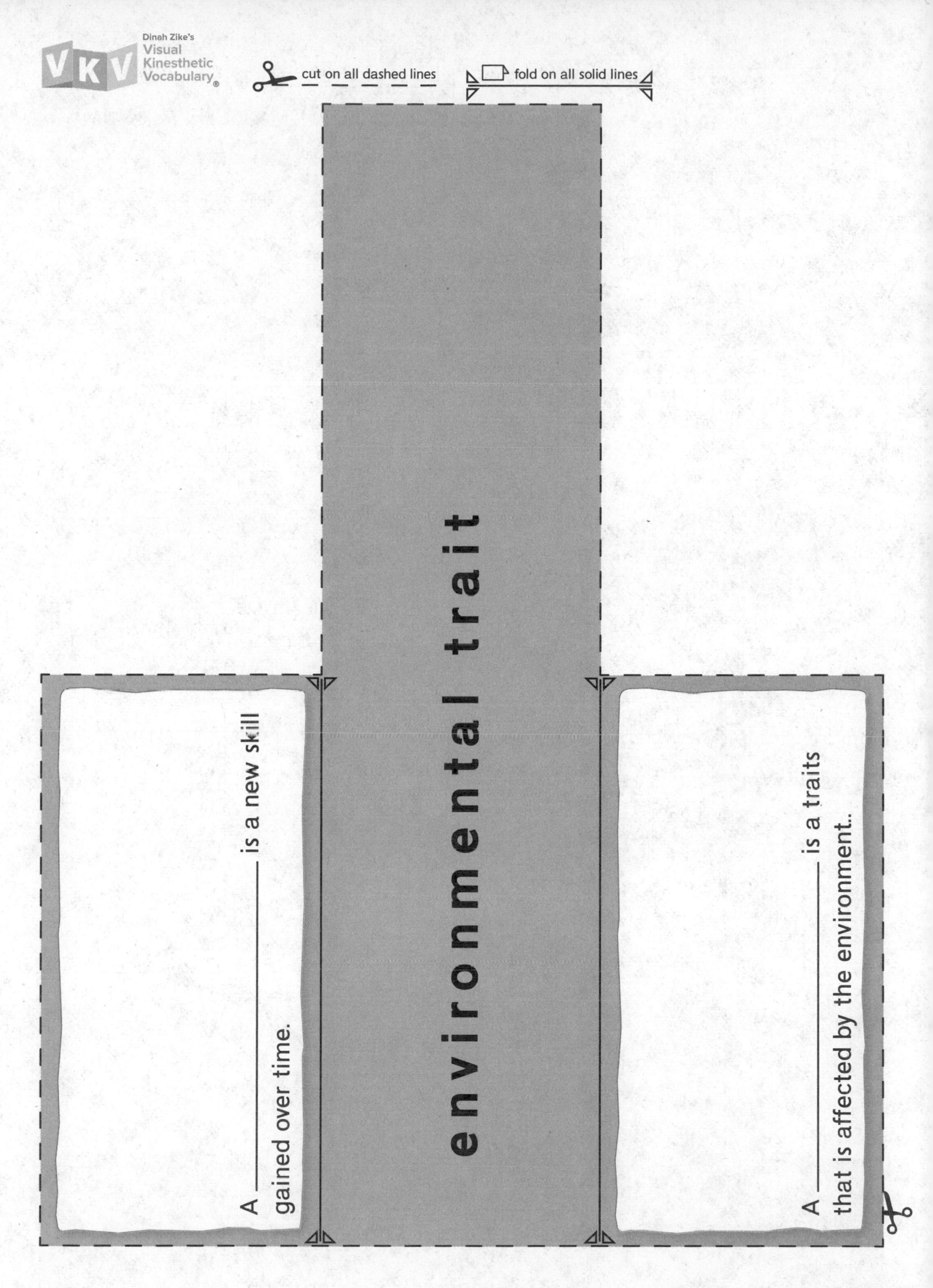
Dinah Zike's
VKV
Visual
Kinesthetic
Vocabulary®
cut on all dashed lines
fold on all solid lines
environmental trait
A ______ is a new skill
gained over time.
A ______ is a traits
that is affected by the environment..

Dinah Zike's Visual Kinesthetic Vocabulary®

cut on all dashed lines

fold on all solid lines

Memory Maker: Draw an example of a learned trait.

learned